LIVE
INTENTION

THE REST WILL FALL INTO PLACE

YOUR GUIDEBOOK TO INTENTIONAL LIVING

CATHERINE EPSTEIN

Reiki Master and Certified Professional Coach

foreword

Her eyes opened at 6 AM, knowing that this day would be different from any other. How different? She wasn't quite sure. She just had a strong feeling that life was about to change.

Her knowledge came from a place deep inside her, and although she didn't know what it meant, she felt surrounded by PEACE.

There she sat in her meditation corner, hearing the sounds of the new day beginning to hum in the background. Picking up her journal and pen, she began to write. The words flowed as if they were channeled from the highest dimensions. Here, she surrendered all of her fears, longings, regrets and hurts. Offering them up to the Divine, she realized: She had a purpose.

This book is the result.

"Out beyond the ideas of wrongdoing and right doing, there is a field. I will meet you there."
Rumi

contents

introduction

The Purpose of this Book

The purpose of this book is to shake you and wake you up to a higher way of living and being in this world. You've lived on autopilot for far too long, and you're learning to live with purpose. You're tired of waking up every morning and "hitting the ground running," going through your day on autopilot, falling back into bed, only to wake up the next morning and do it all over again. Most people live like this until they are hit by what I call the "cosmic 2x4," which stops them in their tracks for a time. It might be an illness, or a challenging family situation, a life-changing moment or one of life's little hiccups. But it is meant to bring you back to yourself. It's a time where you ask yourself: "What's my purpose, why am I here?" And if you're lucky, this event causes you to make changes and not go back on autopilot.

But if you're like most humans, it's hard to maintain the momentum of growth unless you see an immediate payoff.

That's where this book comes in. To help you see the immediate payoff of *living with intention.*

Living a life on purpose. Waking each day in a powerful way, affecting the world around you. Expanding the universe and unifying the hearts of men and women.

That's why you're here.

Don't wait for the "cosmic 2x4." Let this be your sign.

The purpose of this book is to provide a simple framework for living a more intentional life, so you can finally manifest your dreams, be authentic, have healthier relationships, and live more fully from your heart. The instructions and exercises provided will guide you through the simple steps, keeping you grounded and balanced. It will teach you how to overcome the things that have held you back so you can **live with intention**, and watch the rest fall into place.

Begin by asking yourself this question:

"Are you living your highest and best life?"

chapter one

What Is Intentional Living?

"All doubt, despair and fear become insignificant when the intention of life becomes love."
Rumi

To live with intention means to live in alignment with your true nature and purpose. When you think about life, it all comes down to energy -- seen and unseen -- and when we live with intention, that energy aligns and *the rest falls into place.*

When we live with intention, we are existing in a place of pure positive energy, which is our true nature. When we live with intention, things unfold for us in a very different way. People, experiences and lessons show up for us, so that we work through our deepest issues, transform, expand and grow.

The concept of intention is as mysterious as the universe itself, and to bring it down to the order of the mundane world presents many challenges. Trying to explain the concepts of the unlimited universe, the divine mind and unseen forces is like trying to teach the French language to a cat.

While some people feel that the act of setting an intention is a concrete idea or a declaration with visible and tangible results (which can be true), I have a different view.

As a Transformational Life Coach who has helped hundreds of people to master their energy and who has guided them to live more authentically and more fully from their hearts, I've made it my life's work to help people overcome what's been holding them back; to regain the clarity, confidence and courage to live lives of meaning and purpose.

My commitment and passion to spreading the word of intentional living extends beyond myself and the scope of my work. It's been a universal force which has guided me to transform my own life and to reach out to anyone ready to hear this message. It's my calling, what fuels me forward.

So here is my way of defining **intention:**

Intention is the alignment of a desire toward a specific outcome.

I see intention as a more powerful and unseen connector to the *"all-that-is-ness"* of the universe. Like many metaphysical concepts, it's simple and complex at the same time. The power of intention is really what you make of it. **And that is where we begin.**

Let's break it down this way:

ALIGNMENT = Matching the energy in your mind, body and spirit.

DESIRE = The deepest longings of your heart.

OUTCOME = The end result, yielding that which is desired or meant for you.

When we live with intention, it brings a sense of order and clarity into our lives. It is the opposite of living on autopilot. All of our thoughts, beliefs and interactions come from a deeper place within, and our intentions become the guideposts pointing us in the direction of our futures.

Through conscious action, the path becomes clear, and our emotions tell us when we are straying from the path or staying right on course. At our basest level, after our physical, emotional, mental and social needs are met, it is our deepest need to be the best we were born to be.

Yet the conditions of daily living and our individual habits and patterns picked up along the way have challenged us with limiting beliefs. We've adopted these beliefs as our own, and they often operate in the background of our conscious, unconscious and subconscious minds.

There are many challenges in our lives preventing us from living this life the way we were fully meant to live it. To overcome these challenges is the next step towards intentional living.

We'll discuss these challenges (and how to overcome them) in a later chapter, but I'll identify some common examples below:

- **False Beliefs**
- **Self-Deprecation**
- **Victimhood**
- **Procrastination**
- **Fear**
- **Feelings of unworthiness**
- **Perfection**
- **Overwhelm**

Chapter Summary

What is intentional living?

Intentional living is defined as *the alignment of a desire toward a specific outcome.*

Living with intention is the polar opposite of living on autopilot. It is living **on purpose.**

The key to intentional living is having a focused, conscious awareness of your thoughts, feelings and emotions. Living with intention helps us to live with power and purpose.

EXERCISE:

For the next 24 hours, be the observer of your life. Look at everything from a non-judgmental place. Listen to the way you speak to yourself, noting the thoughts that float in and out of your consciousness. Simply practice observing these thoughts. Take notice.

The Four Pillars of Intention

Before we go further down the path of discovering the different ways of intentional living and how to use and manifest the concept of intention in your life, let's first examine the underlying components -- for an understanding of the foundation, built by the four pillars of intention, will help lead you to its core. In the larger view, you'll be able to see all things as energy.

There are four pillars of intention: Alignment, Awareness, Acceptance and Appreciation.

Pillar One: Alignment

On a simple level, alignment is about linking one's head, heart and gut with Universal Source Energy. It is about connecting our physical being to the unseen world, the cosmic consciousness, the *all-that-is-ness* that I mentioned earlier. It is knowing that we exist far beyond these physical bodies and material lives -- our egos, our careers, and our other forms of identity (wife, husband, boss, mother, friend).

Alignment is about lining up *all* of your energy: **inner, outer and higher.**

That's the simplest approach, but as we know, things of this nature aren't always so simple. Part of the human condition is to take a simple concept and make it complicated.

When you begin to harness and apply the power of intention, you can navigate the tribulations of life from a higher perspective. You learn that even though "stuff happens," you'll feel the anchor of alignment serving as your inner source of strength. I know a woman who had experienced twelve deaths of family, friends and acquaintances over two years; plagued by family issues, it was as though tragedy seemed to follow her. Despite her string of misfortunes, she was very clearly able to put her life in perspective, knowing she'd get through the hardship. Her guidance point was her intention of living a peaceful life, to which she remained anchored.

You see, alignment was the key. While it couldn't change her circumstances, her intention kept her going.

Conversely, we all know people who complain about everything and love to play the victim role at every turn.

This is not about ignoring what is happening in your life, nor is it an example of *"spiritual bypassing"* (a term used for people who think everything is light and love, without doing the real work of accepting their feelings).

It is simply about understanding the laws of the universe, and living in a higher way. When you think about it, there will always be challenges and lessons in life, but what matters is how we deal with them. Staying in alignment is our #1 principle, for living without that connection to our higher power/inner source/whatever you like to call it is like trying to run the vacuum cleaner without plugging it in. We can run through the motions of vacuuming up the floor, but it really won't do the job.

Pillar Two: Awareness

Awareness is a multi-pronged, mindful approach that requires being conscious of all things that are currently happening in your energy field at any given moment. You'll become more astute at detecting your emotions as you begin to deliberately set the intentions which create your life.

By becoming more aware of your thoughts, feelings and emotions, you'll be able to step up your game.

Let's discuss awareness like this: Bring your attention to your entire being right now. Perform a brief check-in with yourself, head to toe. What are you feeling? What's your mood? Are you still holding on to some triggers from your past?

Awareness acts as a barometer for us to focus on our inner world, our outer world, and our higher world. It lets us know whether or not we are in alignment.

But the full concept of awareness includes remaining objective -- even if in practice, we don't like what we find. Identifying the feeling is the first step toward mastery. Saying "I feel anxious," has a different energy and result than saying "I am anxious." It is not a denial of what we are feeling, but it is a distancing of that feeling from our identity. It's simply a non-judgmental way of acknowledging our feelings in the present moment. And as we know, feelings can be fleeting when we acknowledge them, release them and move on. It is when we deny them and try to tamper them down that they inevitably erupt elsewhere in our lives, overcoming and overwhelming us.

When we operate with a fuller awareness, it means operating less on "autopilot," and more on "inner pilot."

Three Ways to Strengthen Awareness:

1. **Stop** in the moment, and do a check-in. What am I feeling? What do I need?
2. **Take breaks** throughout your day to sit in silence. Meditate on the stillness within, even just for five minutes.
3. **Be present** and focused -- while standing in line, while in conversation, while driving, while eating. Keep an awareness of your five senses during each experience.

Again, objectivity is key here. Be very compassionate with yourself.

Pillar Three: Acceptance

Acceptance can be tricky for some, because they think it means that they need to tolerate everything -- even that which is negative. For our purposes, we will discuss what it means to be in acceptance. The deeper meaning is to accept that which is present, and decide from there whether to keep it in your life or not.

Said another way: When we spill the milk, do we just stand there and yell, "Why me?" We could waste a whole lot of energy on blaming, shaming or yelling. Or we can take a breath, accept, and clean it up.

Let's pick a different, deeper issue: Say your husband of thirty years comes to you and says he's leaving. He hasn't been happy, and he wants a divorce. If you could view this from a higher perspective, you could understand that the relationship has served its purpose and run its course. **Acceptance** of the situation helps one to make peace with it much more quickly. This is not to say you should ignore, or close yourself off to, the emotions you feel related to the event. But you must make peace with what is.

Humans are blessed with a range of emotions, from deep sadness to pure bliss and everything in between. So much of the beauty in our earthly experience is to feel the full range of these emotions.

Part of the acceptance piece is to acknowledge the present emotion, and move through it consciously. In the case of the divorce, there will be many layers and stages of grief to process, but when we do that consciously, we have better control over how we dwell in that particular stage. Once again, this isn't denial or spiritual bypassing, but a mindful approach to living more intentionally.

If your intention is to live a life of peace and happiness, the Universe will bring up anything in your life that is not aligned with that intention -- you can be sure that an unhappy marriage will not remain in your experience.

Another example may be one you are familiar with: You set an intention to bring more money into your life, by asking for abundance. What happens sometimes is that you receive an unexpected bill, your car breaks down, or your rent is raised. This is just another instance of the Universe bringing up that which is ready to be healed -- in this case, *scarcity consciousness*. When you ask for more of something, it is implied that you don't have enough. Your focus is on "not enough money," instead of gratitude for what you do have, and to that, the universe responds.

From here, there are two ways these scenarios can play out:

1. **Remain stuck** in scarcity, get upset and complain about how things always go wrong (attracting more negativity), *or...*
2. **Affirm** that your intention is to attract prosperity, and that you are grateful for the solution that is already on its way. Focus on the abundance that is already present in your life, in every way.*

**You must do this, even if your bank account is low!*

Pillar Four: Appreciation

When we live with appreciation for *all* that is present, it creates the space and conduit for more positivity to flow into our experience.

Deep appreciation lets us step out of our heads, allowing us to connect our hearts with the expansive energy of the Universe. Maintain a state of appreciation for life, as it helps us to see through the same lens which Spirit sees us and our lives. When we live this way, we know that we're powered by something beyond these physical bodies.

We have the entire Universe guiding the way, cheering us on -- that is what fuels us.

Can you sense the power behind appreciation and gratitude? Appreciate your breath, your life, your body (the complex machinery that it is, and all that it allows you to do), your home, your comfortable bed, your family (*yes, even your family*!). Find beauty everywhere: the look on a person's face as they hurry to the subway, the bird landing in the tree outside your window, the way your dog greets you when you return home.

chapter two

When we fill our hearts with appreciation, it is the highest form of communing with the Universe. It is the form of expression that fills us with *pure joy.* Find that place within you, and tap it into it often. For inspiration, practice this mantra of appreciation:

Take a moment now to place your hand on your heart, feeling the energy flow through and give thanks. Tell your heart:

I appreciate you. I appreciate how you've guided me, protected me and how you've carried my hurts, disappointments and grief so well. I live in deep gratitude for you, as we've been through so much together. My intention is to honor YOU, my heart, and I move forward with the intention to live life more fully by my path, and to trust you to light my way.

I feel it. I know it. I LIVE it.

The energy of appreciation is expansive -- it allows us to view the world from a higher perspective -- to see beauty in all faces, places and graces. It helps us to tap into the unlimited, boundless ocean of beneficial life force **energy**.

Appreciation is the key that unlocks the door to freedom.

Chapter Summary

The four pillars of intention are *alignment*, *awareness*, *acceptance* and *appreciation*. They make up the foundation of living intentionally and help provide the framework for us to navigate whatever life throws at us. When we understand these concepts and begin to apply them more deliberately in our lives, there's a noticeable shift.

EXERCISE:

Choose one pillar at a time, and focus on understanding it and applying it more consciously in your life.

First, write down the pillar you've chosen on a piece of paper.

At various intervals throughout the day, look at the word and remind yourself to embody it.

Check in with yourself throughout the day to keep focused.

Note in your journal at the day's end how this exercise worked for you.

chapter three

Overcoming Challenges, Obstacles & Self-Sabotage

In chapter one, I mentioned how unconscious and subconscious beliefs and patterns crop up and can often stop us in our tracks. Think back to when you were a kid: were you told to keep quiet, and not be a show-off? How has that conditioned your behavior now? Very often, these coping skills acquired early on helped to keep us "safe," but no longer serve us as adults.

When we can understand the mechanisms of these beliefs, we can begin to thank them for being valuable at one time in our lives, but they are no longer necessary. I've seen time and time again how liberating it can be to finally let an old belief or thought pattern go.

The following challenges which many of us face are actually interconnected, and are deeply woven into our psyche. The good news is that we can identify them, bless them and release them for good.

COMMON CHALLENGES	
• **False Beliefs**	• **Fear**
• **Self-Deprecation**	• **Feelings of unworthiness**
• **Victimhood**	• **Perfection**
• **Procrastination**	• **Overwhelm**

By making sense of all the human foibles that come into play, we can focus instead on expansion and living a higher, better life. During the course of our lives, we are constantly bombarded by external forces -- such as the media, telling us that we are not good enough as we are. We need to be younger, taller, slimmer, faster, more beautiful, better dressed.

chapter three

Our own family values instill certain traits within us and sometimes, we adopt these traits as our own (perfectionism, people pleasing, obligation, over-achieving... the list goes on and on). We become a conglomerate of others' needs and expectations, bypassing our own intuitions and becoming a fraction of our true selves.

Don't despair: This isn't meant to discourage you, it's meant to open your eyes to the nature of living the human experience, all the while forgetting the essence of ourselves and "reality."

So many people I meet are generally happy people, with a vague sense of something missing. They like their lives, but feel that they're not living their higher purpose. They say they're happy, but there is a sense that they're settling, merely content -- they *know* there's something more. They make excuses for not implementing change:

"When the kids move out..."
"When I get a better job..."
"When I lose that ten pounds..."
"When I find the perfect partner..."

Our challenges lie in the undercurrent of our psyches, such as the belief that we're not good enough, that we'll never have a healthy relationship, that we're no good at math or science, that we'll never make it as a musician, that we should play it safe, that we should settle for someone who'll take care of us.

So many voices... and they're *not yours!*

When you begin to work with the four pillars addressed in chapter one, I assure you that you will have some eye-opening insights as to where some of these beliefs originate. It's not just ourselves or our families, it's much larger than that: our communities, our societies, our long-instilled beliefs handed down through time for generations. When you begin to face your own set of challenges, you'll be amazed at how freeing it is to overcome them.

You're not broken and there's nothing to fix.

Let that sink in.

chapter three

If you're reading this, I know you're the kind of person who wants a deeper understanding of yourself, and that you're making it a priority to live a life of purpose -- to live up to your full potential, and make a difference in your life, if not the world. Coming from that perspective, you'll be able to break through what's been holding you back, and move forward with freedom and joy.

To overcome these challenges, you must view your behavior through a lens of compassion and understanding. Begin practicing this in your daily life: Bring your awareness back to the fact that you are energy, and you are so much more than the physical self. Start there, and know that when you face resistance, you can take the steps to acknowledge those internal walls and move on.

Beneath all of these challenges lies fear, and that is where the work begins.

Overcoming the fear of anything starts with recognizing this:

You are bigger than your fears.

Chapter Summary

Acknowledging the challenges that crop up as we grow and expand into our true selves is the next step in living with intention. These challenges or roadblocks have been with us for such a long time that while they may have become part of our personality, they do not define us. Overcoming these challenges in a healthy, compassionate way is our main task.

EXERCISE:

Try these steps.

First, take your power back by recognizing that you're stronger than you think.

Then, acknowledge the fear. You can even thank it, for what it will teach you.

Access the power of the Universe. Tap into that. Trust it.

Focus your thoughts on the highest outcome.

Make peace with where you are.

The Guide: 10 Steps to Live an Intentional Life

In this chapter, I will outline the ten steps for you to live a more intentional life. This guide will help you to stay connected to your energy, and will help you to navigate the ups and downs of life. Living **on purpose** doesn't mean there will always be smooth sailing, it just helps to navigate the waves in an easier way.

When the waters get choppy, use this guide to help come back to the center of your power.

TEN STEPS TO INTENTIONAL LIVING

Step One: Make peace with where you are.
Let go of the past, release all those things and events which played a part in your growth. Making peace doesn't mean forgetting, it just means no longer holding onto the energy that keeps you stuck.

Step Two: Take full responsibility for yourself.
Let go of victim or blame -- its lower energy will keep you playing the part over and over until you decide to let go. Take responsibility for who you are, and who you are becoming.

Step Three: Learn to live in the present moment.
This is where our true power is. There is only one moment, and that is **now.**

Step Four: Align with your highest and best self.
Acknowledge that you are a being of energy and that you're so much more than this physical body. Keep bringing your awareness back to that.

Step Five: Take back your power, and carry yourself with confidence.
Notice when your energy is directed into fear, worry or other draining emotions. Call it back as soon as you recognize it. Draw it back into your core (your center, or your solar plexus).

Step Six: Speak to yourself with kindness.
Notice the way you speak to yourself -- adopt a softer, kinder attitude. Treating yourself with love and compassion is the only way forward. Learn to speak to yourself as you would when encouraging a small child to learn something new.

Step Seven: Reframe your beliefs.
As part of living life, we have all picked up thoughts and beliefs along the way. Most aren't true, and usually have come from somewhere or someone else on our journey. Notice what beliefs you hold ("I'm not good at... math, singing, art), and begin to tell yourself a new story.

Step Eight: Receive all compliments.
Most of us are quite adept at deflecting compliments:

"Your hair looks great!"
"Oh, I'm just lucky today."

"I LOVE that dress."
"This old thing? I just pulled it out of the back of the closet, I've had it for years."

Sound familiar? Begin to notice your tendency to deflect any compliments -- it's tied directly with our ability and openness to receive what the universe has in store for us.

Step Nine: Embrace your fears.
Know that you are bigger than your fears. Acknowledge them, bless them, thank them and move on.

Step Ten: Be who you are meant to be.
Live your life on purpose. Always embody and know that you are a being of wisdom, and that is your true nature.

Chapter Summary

This chapter is all about taking steps toward intentional living, breaking through the old patterns that have held us back. You may know this as the "comfort zone." Change isn't easy and sometimes, it's in our nature to continue on in the ways that are most familiar to us.

How many of us play it safe as we listen to the nagging voices inside? The voices that tell us to stay small, or quiet; that we might not be good enough, we might not succeed, we might make a mistake. Well we've all been there. And it is only human to stay with what is safe. How do you know when it's time to follow your heart? Part of learning about yourself and the inner workings of your heart is a very deep process -- but it's also one that can be very rewarding, and helps you to shed the layers of other voices which have accumulated over lifetimes.

What is it that pulls at your heartstrings? How can you get out of the box, so neatly wrapped, that is your life?

Let your heart lead the way, and follow the path with your emotions as your guide. Check in with your inner self often.

EXERCISE:

Do one thing outside of your comfort zone. Every day. Stretch your comfort zone with small steps -- different types of classes, different routes to work, new lessons, a home business. Approach each new endeavor as an adventure, joyful like a child.

Are you connecting with joy in your life? Are you sharing your light to bring joy to others?

chapter five

The Process

In this chapter, I will show you a very simple process to help guide your every day self-work. This is very important to incorporate into your routine, as it sends a message to your subconscious that you are worthy, that you are in the process of becoming the very best version of you, consistently on the path of improvement.

The combination of touch, breath work and alignment of your energy will carry you far beyond the ordinary, and into the realm of extraordinary.

APPRECIATE

Every day when you wake up, connect with your inner self by placing one hand on your heart and one hand on your belly. Take three deep breaths. Express your gratitude.

INTEND

State your intention out loud: "I intend to live in alignment with my true soul purpose," "I intend to live as my higher, best self." Create your own, which speaks to you and your path.

FOCUS

Throughout the day, notice your thoughts and what you are focused on. Write them down if necessary. When you find that you're feeling frustrated, down, or any other emotion not aligned with your intention, begin again by restating your intention. Sometimes it helps to take a break, go for a walk, pet your puppy or kitty, anything to disrupt the downhill pattern.

Learn to pay attention to all that you are feeling, and allow it. Name it (if you can). Sit with it, give it space. Allow it to be. Then, return to step four, bring your awareness back into the moment, and restate your intention.

chapter five

FEEL

What does it feel to live as your higher self, your true soul purpose? Use your imagination, and act as if you're already there. View the world and all of your interactions as your inner-being would. When you try on this future self, it's very empowering. You walk straighter, the world looks clearer, and suddenly, you see things through very different eyes.

Chapter Summary

The process is to get you to connect with your body, and begin to develop a short, doable daily routine to live as your higher self. You are actually involving the body, mind and spirit, connecting mindfulness to your physical self, aligning with the Universe. By incorporating a daily routine, it becomes a habit that will be as ordinary and commonplace as brushing your teeth. This process helps to integrate all of these teachings at a deep level. You'll begin to know when you are out of balance, and can quickly bring yourself back into alignment.

EXERCISE:

Set a day to begin -- put it in your calendar. Keep a note by your bed with the five steps from above, and set your intention that when you wake up in the morning, you will begin the process.

Practice each day until it becomes second nature. Upon awakening, these steps will be automatic.

Wrapping It Up with the Coach

Let's pause and take a breath! We've covered a lot of ground.

As I said at the beginning of this book, intentional living is both a simple and complex topic. We fundamentally understand that we want to live a life on purpose and not on autopilot.

Yet in the human existence -- especially in our fast-paced society -- this becomes a difficult task. It's certainly easier to live on autopilot (yes, I admit it!) but I also like to say that there's no comfort in the comfort zone.

It produces a vague uneasiness that hides in the background and leaks out in other ways. Sometimes it manifests as depression, overwork or emotional distance.

Intentional living is about being connected to the universe, and to have an understanding of all the energy in and around us. Knowing that we are here to learn, to grow and to expand our consciousness.

We know we are beings of energy, here to play on the earthly plane and learn to live a life in which we consciously create the highest, best experience for ourselves. It's time we shake off the sleepiness and the complacency, and learn to step into our highest functions. We are here to create our best lives, to affect the world all around us and to live in the highest states of love, abundance and *joy*.

It's time for you to stop playing small: the universe is calling you, your heart is calling you and there is *no one else who can be you but you.* It may sound trite, but you've been given a special purpose in this world -- and you'll know what that is by getting clear and listening to your heart.

It is my hope that you've gained an understanding of how to live a more intentional life so you can become who you were born to be.

chapter six

In the course of my work, I've helped hundreds of people to people to gain mastery over their energy so they can manifest their dreams.

We begin first by setting the intention to live a life on purpose. In this book, you've been given a framework and instructions for coming into your full power as a creative being on this earth. Feel the call to share your gifts with the world, simply by being yourself.

Answer the call to live with intention, and the rest will fall into place.

Alignment, awareness, acceptance and appreciation.

Be grateful for where you are right now. Intend to live as your highest self. Focus on your thoughts. Feel as if you are already living your best self.

THIS SEVEN DAY WORKBOOK

is intended as an exercise in self-reflection, serving as a guidepost to express your deepest desires and set you on the path to living with intention. Every day, when you sit down to answer each question, let the answers flow from your heart (not your mind). Imagine your alignment in mind, body and spirit, and let it flow from within you.

DAY ONE

What is your intention for your life?

DAY TWO

What is standing in the way of your happiness?

DAY THREE

What are you afraid of?

DAY FOUR

What does your life look like in one year from now? Five years?

DAY FIVE

What has been the biggest gift in your life?

DAY SIX

When do you feel most in tune with yourself?

DAY SEVEN

What brings you JOY?

ABOUT THE AUTHOR

Author Catherine Epstein, founder of the Living Lotus Group and Living Lotus Coaching, is an experienced guide and instructor in meditation and mindfulness practices as well as an author of The Divine Dining Method, Reiki Master of the Traditional USUI Method, Graduate Gemologist of the Gemological Institute of America (GG), and Certified Transformational Life Coach (CPC). A mindfulness mentor, and speaker – Catherine's specialty is helping people to live more fully from their hearts, with an approach to coaching that's part laser-beam focus with a bit of Sagittarius optimism. Read more about Catherine at www.livinglotusgroup.com.

ADDITIONAL WORKS